COPYRIGHT

Bannon River Books, LLC
www.BannonRiverBooks.com
Rutland, VT

For licensing requests, other permissions, and bulk orders, contact the publisher at:
info@BannonRiverBooks.com

For free book activities, author visits, and sustainability resources, contact the author at:
www.TheCarbonEmissionsProject.com

This is a work of non-fiction. While every effort has been made to ensure the accuracy of the information contained herein, the author and publisher assume no responsibility for errors, inaccuracies, omissions, or any consequences resulting from the use of this information.

Cover design by Arlene Sotto, Intricate Designs

Cover art by Luna Hansen

Library of Congress Control Number: 2025910440

Softcover ISBN: 979-8-9855503-5-1
Hardcover ISBN: 979-8-9855503-6-8
Digital ISBN: 979-8-9855503-7-5

First Edition

Printed in the United States of America

a kid's guide to
CLIMATE ACTIVISM

Kimberly Haidinger

Luna Hansen

Bannon River
BOOKS

Rutland, VT

WELCOME

There are people in the world who allow themselves to be <u>marveled</u> by nature every day. My daughter, Luna - the <u>illustrator</u>, is one of those people. She sees shapes in trees and clouds, hears the callings of different birds, and stops to smell every flower. She protects bees, spiders, and small crabs crawling on the beach, enjoys birds in flight, and is <u>enchanted</u> during long walks in the forest. I am lucky to have Luna and her <u>insights</u>. She is my <u>inspiration</u> and the reason for this book.

Luna and I are so happy that you have picked up this book! We hope that it <u>inspires</u> a new relationship for you with the world, creates new <u>climate-beneficial</u> daily habits, and that you learn new important words along the way.

Our beautiful world, and all the nature it contains, needs our help. Your help! Every action - big and small - counts and adds to our <u>collective</u> effort. This book has been created out of a love of nature and our world, and to build a community that takes action every day to protect it. We are grateful that you are joining our <u>climate activism team</u>. Enjoy the journey and share it with friends!

With Gratitude,

Kimberly and Luna

How to have the best experience reading this book:

- We encourage you to read this book with your friends and family so that you can share ideas, exchange and develop thoughts, and organize actions.

- Read it over a couple of days so you can take your time to discuss the different words and concepts.

- We hope this book helps you to grow your vocabulary, especially those words related to <u>climate</u>. We have dedicated a page to presenting each of our big concept words. Additionally, all the underlined words throughout the book are defined in the glossary section at the back.

- If you are not sure what an underlined word means, flip to the glossary at the back and look for that word with its definition.

- With some of the bigger words, we have included pronunciation help in parentheses underneath the word so you can practice saying it correctly while reading.

- Keep your journal or a piece of paper with you while you read so that you can jot down ideas and notes.

ANTHROPOGENIC: means caused by human beings.
(an-thruh-puh-jen-ik)

Scientists agree that climate change is caused by human activities such as burning fossil fuels, deforestation, and agricultural activities.

If human actions caused the problems, it is a very reasonable conclusion that we as humans can solve the problems.

MY leaves Gift YOU air!

6

It can be scary when it feels like there is no solution, or if you do not know how to fix something.

While it is true that <u>anthropogenic</u> means human-caused, it also tells us that we can cause change for good!

If we change our habits, we can have a positive <u>effect</u> on <u>climate</u> change instead of a negative one.

Together we can change the world!

7

VALUES: are the beliefs that determine what is most important.

When we think about our values, we should think about how they impact not only us, but the people and world around us.

Those values can be the beginning of good decisions for the Earth and everything that lives here.

Examples of values are:

Respect

Responsibility

Kindness

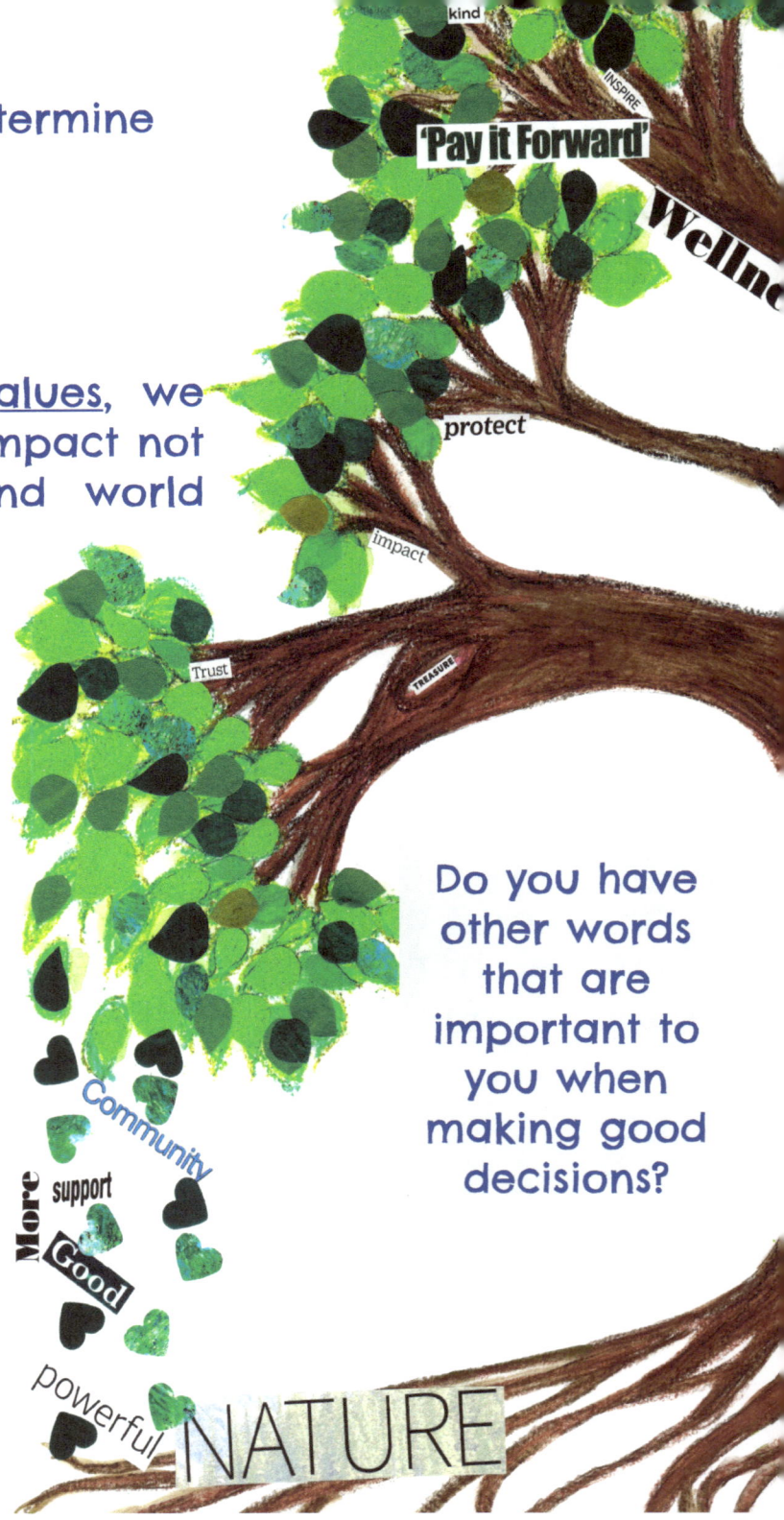

Do you have other words that are important to you when making good decisions?

kind

INSPIRE

'Pay it Forward'

Wellne

protect

impact

Trust

TREASURE

Community

More

support

Good

powerful

NATURE

8

FRIENDLY

Your Forever Home

help

SWEET

Thankful

HEALTH

Sometimes we compromise our <u>values</u> for <u>convenience</u>.

That means instead of doing what is right, we do what is easiest.

courageous

Life

We can choose <u>values</u> that have a good impact, even if it isn't the easiest or fastest thing to do.

More

<u>Love</u>

legacy

You matter

CONVENIENCE:

(kuhn-veen-yuhns)

is the thing that is quickest and easiest but, often in relation to the environment, not the right thing to do.

That means convenience can sacrifice our values.

Read each bubble for an example!

Throwing away old toys instead of finding them a new home or use.

Driving instead of walking or biking.

Using underline{single-use} plastic/paper instead of reusable items.

Dropping garbage on the ground instead of throwing it out.

Choosing plastic water bottles instead of reusable ones.

Throwing out food instead of _composting_ it.

Can you think of times when you've done these things? Can you make a better choice next time?

The carbon footprint of food:
(kahr-buhn)

The word "carbon" refers to all the gases that are emitted during the production and distribution of food.

The word "footprint" is used to help give the idea of leaving a mark, just like your foot in the sand at the beach. Food leaves a mark on the climate.

The word "food" is used to describe everything from the creation of the ingredients to getting the food on your plate to disposing of food waste that you do not eat.

NOTE:
The carbon footprint of food can also be called food carbon emission.

When you think about the impact of your food on the <u>climate</u>, you have to consider the process that your food has to go through. Each of the steps listed on the next page adds to your <u>food carbon emissions.</u>

When you turn the page:

- Take a minute to read through all the steps.

- Can you think of more examples that add to the <u>carbon footprint</u> of your food?

- Can you think of anything that can lower the <u>carbon footprint</u> of your food?

Global impact. FARM TO Consumer Groceries RESTAURANT everyday framework MEAL-PREP TO AND FROM Convenient shopping

sweet carrots Like carrots need carrots EAT carrots

Each step of the food process is listed with one example next to it.

- Land usage = land used to make our food instead of being, for example, a forest.

- Farming = to grow our food, or food for our food (think of feeding the cows, pigs, and chickens).

- Processing = harvesting the crops is an example.

- Transporting = by truck, ship, and/or plane to a local store.

- Packaging = putting it in a bag or container to prepare it for sale.

- Handling = stocking it onto the grocery store shelves.

- Cooking = the food at home, at school, or in a restaurant.

- Throwing away = uneaten food.

Can you think of other examples that could happen at each step?

How do you think a box of cereal gets into your bowl?

14

Because of all those steps, all food types produce a different amount of carbon.

● An Example of Food Carbon Emissions by Food Type

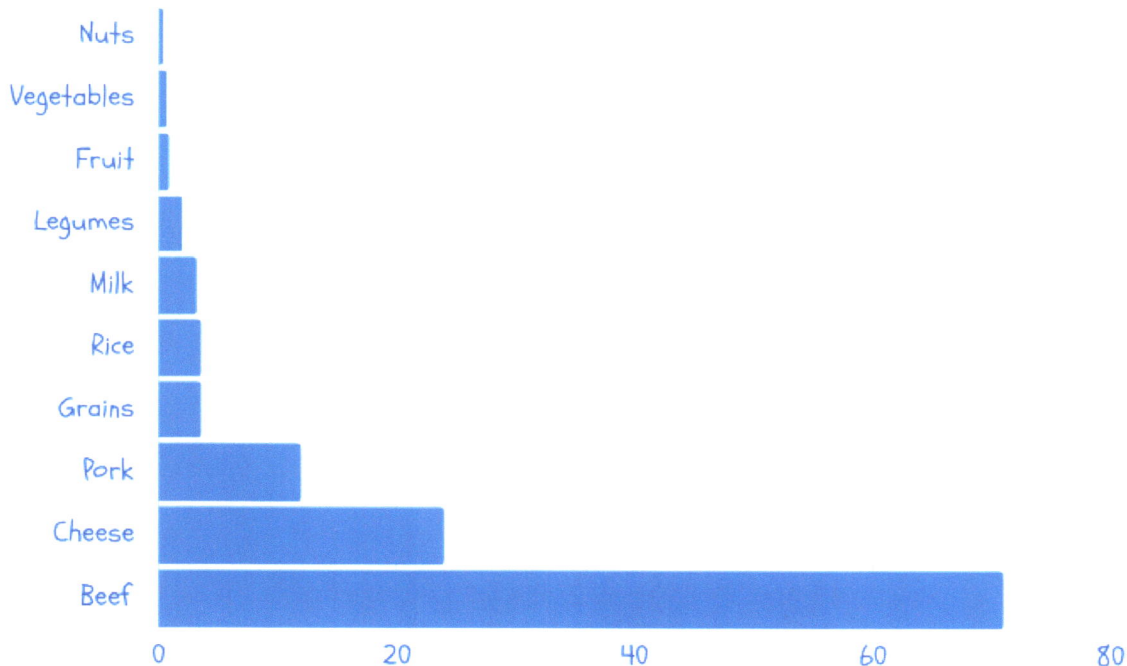

Approximate Kilograms of Carbon Emitted by the Production of Food

Graph source: www.un.org/en/climatechange/science/climate-issues/food

- Food is responsible for approximately 25 percent of all greenhouse gas emissions (GHG).

- Looking at the graph, can you think of a way to lower your food carbon footprint?

- What are your favorite low-carbon foods to eat?

15

FOOD WASTE: is food that is not eaten and gets thrown in the garbage.

Food waste is very heavy and takes up a lot of space!

When food waste is mixed with regular garbage, it rots in a landfill and emits methane, which is one of the greenhouse gases (GHG).

It is really important to separate food waste to recycle or compost it instead of adding it into the trash.

COMPOSTING & FOOD RECYCLING:
(kom-pohst-ing) & (food) (ree-sahy-kuhl-ing)

Composting is a process that turns our food waste into a material that can be used to enrich the health of the soil in our garden. It is a little bit of work and might be less convenient, but composting helps to reinforce our good climate values.

Some towns have a service and can pick up your food waste from your house, and compost it for you.

Maybe you can check and see if your town or a local company offers that as a solution.

Or if you have a garden at home, maybe you can try composting!

17

SINGLE-USE ITEM: describes something that you use only once and then throw away.

It means we are adding to the garbage problem every time we use a <u>single-use</u> item.

Examples of items that get used just once:

- plastic wrap

- aluminum foil

- takeaway food containers

- plastic to-go drink cups

- straws

- paper towels

- gift wrapping paper and bows

Can you <u>brainstorm</u> some substitutions for <u>single-use</u> items that you use?

CONSUMERISM:
(Kuhn-soo-muh-riz-uhm)

is buying more and more things that you may not really need.

The problem:

Sometimes, the moment we see something, we get the desire to have it immediately, but afterwards realize that we do not really need it.

Maybe it is a cute stuffed animal, or a fun-looking new toy, sporting equipment, a pretty dress, or hair clip.

Every time we buy something new that we don't really need, something we already have goes unused and becomes waste.

20

The Solution:

Reduce! Maybe we do not need to buy everything.

Reuse! Maybe we can exchange toys and clothes with friends so that we have new things, but do not have to buy new things. We can also repair the things we own.

Refuse! This might be the most difficult solution to get used to, but maybe we have to practice just saying no to purchasing new items.

Can you think of some examples of when you reduced, reused, or refused?

FINITE: means having a limit, like the size of our planet. (fahy-nahyt)

Planet Earth does not get bigger just because there are more people and things in the world.

Just like the finite size of your bedroom. You can only have so many toys, pieces of furniture, and clothing.

You probably cannot hide a hundred huge boxer puppies in your bedroom because your room has a limit on how many can fit inside. It has a finite amount of space.

What other things would be impossible to hide in your room because of its finite space?

BIODIVERSITY: "bio" means life, while "diversity" means different or <u>containing</u> variety.

The diversity of life throughout the world and especially in a particular habitat is important because the variety of plants and animals keeps the world healthy and strong.

All living things work together. This means that the more varied species of plants and animals living in one area, the more jobs that get done.

This amazing diversity makes our world better able to <u>adapt</u> to <u>environmental</u> changes.

An Earth with more <u>biodiversity</u> is greater able to fight climate <u>change</u>.

Can you name some of the things that make the seashore <u>biodiverse?</u>

How about a forest?

WHAT'S THE DIFFERENCE BETWEEN WEATHER...

Weather describes small temporary changes, like rain or snow, happening in a small area, like your town or state.

While the word climate refers to the less specific and more long-term atmospheric trends of a large region.

...AND CLIMATE?

Extreme weather can be a sign of <u>climate</u> change, but it is not the whole picture. <u>Climate</u> change refers to the planet's temperature increasing. That increase in temperature causes other <u>effects</u> like:

the warming of the oceans, wildfires, flooding, and melting ice.

All these things affect wildlife and humans in ways that are very dangerous.

CARBON EMISSIONS:
(ih-mish-uhnz)

the word "carbon" refers to one of the most basic elements in all living things.

The word "emissions" means something that has been released into the world.

A little review:

Carbon is natural and even essential to life on earth. It is not actually harmful on it's own, until - because of human activities - it becomes too abundant in the atmosphere.

A lot of people think that carbon comes only from dinosaur fossils. That is not true, but carbon did begin to form a very long time ago - maybe even billions of years ago! That's a very long time.

Carbon is made from plants that decayed in bogs and swamps, as well as plankton from the bottom of the ocean floor.

Today's coal was a plant in a bog over 300 million years ago, and oil and gas were plankton 1500 years ago.

FOSSIL FUELS: form far beneath the Earth's surface from the carbon-rich organic matter of dead plants and animals.

Today, when we talk about fossil fuels, we are referring to the energy source that we get from that decayed material.

Specifically, coal, petroleum (oil), and natural gas.

A little review:

Around 4000 years ago, humans realized that coal could be burned to create heat. Today, we rely greatly on fossil fuels (coal, oil, gas) for energy.

We drill and mine fossil fuels to heat our houses, fuel our vehicles, and run the electricity at home, school, and work.

Unfortunately, when we burn fossil fuels, they release greenhouse gases (GHG) into the air.

GREENHOUSE GASES (GHG): refer to any gas in the atmosphere that traps heat.

The name - greenhouse gases (GHG) - makes sense because the atmosphere is made up of gases, and it does the job of a greenhouse. There are some gases that occur naturally, but the term greenhouse gases (GHG) has come to be used to describe the harmful increase in gases due to human activities.

Earth's atmosphere - which is made up of gases - acts like the glass of a greenhouse. It allows the sun's rays through but prevents the sun's heat from escaping. This is called the greenhouse effect, and it is what keeps the Earth warm enough for us to live on.

The gases that we talk about the most are carbon dioxide, methane, and nitrous oxide.

GLOBAL WARMING: is just like it sounds. Earth is getting hotter and hotter.

Our reliance on <u>fossil fuels</u> has increased the concentration of <u>greenhouse gases (GHG)</u>.

Too much of these gases increases the Earth's temperature because more heat is trapped close to the Earth's surface.

The warmer temperatures are causing all the changes to our <u>weather</u>.

Let's put it all together!

1. Plants and animals decay far beneath the Earth's surface.

2. Carbon is transformed, <u>fossilized</u>.

3. Coal, gas, oil: the <u>fossil fuels</u> are created.

4. Humans drill for <u>fossil fuels</u>.

5. Humans burn <u>fossil fuels</u> for energy.

6. Gases are released into the <u>atmosphere</u>.

7. Gases build up and create too much of the <u>greenhouse gas effect</u>.

8. The planet heats up, and we call it <u>global warming</u>.

9. The hotter temperature creates changes in the <u>weather</u> patterns, and we call that <u>climate</u> change.

10. Now we organize to reverse the damage and lessen the build-up of <u>greenhouse gases (GHG)</u>. We fight <u>climate</u> change!

**Millions of years
of science
summarized into
10 easy steps!**

CARBON SINKS: are our best friends in the fight against <u>climate</u> change. <u>Carbon sinks</u> are our forests, oceans, and soil.

<u>Carbon sinks</u> <u>absorb</u> more <u>carbon dioxide</u> from the <u>atmosphere</u> than they release, so they are important tools in helping to remove <u>greenhouse gases (GHG)</u> from the <u>atmosphere.</u>

The Earth's greatest <u>carbon sink</u> is the ocean. It stores 38,000 <u>gigatons</u> of <u>carbon.</u>

That is a really big number!

Our oceans are so important in the fight against <u>climate</u> change!

CARBON NEUTRAL AND CLIMATE NEUTRAL

The word carbon is most often referred to in discussions about climate, however, it is more accurate to aim to be climate-neutral rather than just carbon neutral. In both cases, the idea is to have an equal balance between emissions and absorption or removals.

Here are some ideas that will bring you closer to becoming climate-neutral:

You can use solar panels to provide all the energy that you use in your house.

You can plant a tree to replace the tree that you cut down for Christmas.

You can shop for "new" clothes at a second-hand shop.

You can use recycled paper for your craft projects or cut up old fabric to make a new dress.

Can you think of more ways to be climate-neutral?

CLIMATE OVERSHOOT: can happen because of <u>climate</u> change. It means the earth has gotten hotter and even when we work to fix it and the earth cools back down (or is <u>stabilized</u>), the <u>effects</u> of <u>climate</u> change are still felt.

Here is an example: Have you ever burned the roof of your mouth with hot pizza or apple pie? Food remains too hot right after it is taken out of the oven. It is a good idea to wait until it cools down before you eat it.

This is your pizza's or apple pie's overshoot.

If we are successful in decreasing <u>global warming</u>, we will still experience the <u>effects</u> of <u>climate</u> change while the planet recovers and cools back down.

That doesn't mean we shouldn't try! It does mean we need to try harder to lessen the impact.

CLIMATE ACTIVIST: describes anyone working to protect the planet's health.

Save our Oceans

Climate describes our global weather and is important because the pattern of our weather is changing. It is not about today's weather or tomorrow's weather, but about the changes in the pattern of the weather we experience.

SAY NO To PLASTIC BAGS

Activists believe strongly enough to take action and get involved in activities to bring about change.

There is No PLANET B

Can you brainstorm some ideas about what actions you can take to get started?

(Hint: Keep reading, there are some helpful ideas coming.)

STOP GLOBAL WARMING

How do all these words come together?

All the people in the world need to work <u>collectively</u> to reduce the <u>emission</u> of <u>greenhouse gases (GHG)</u> and stop <u>global warming</u>.

Governments, scientists, and businesses all around the world are working on solutions, but every individual has a part to play, too.

The <u>conveniences</u> that have become a part of our daily habits have to be changed to place our <u>values</u> for the planet above what is easiest. Our grandmothers, and great grandmothers, and great-great grandmothers were careful not to create <u>food waste</u>, they lived without <u>single-use</u> items, and the abundance of <u>consumerism</u> that many of us enjoy living with today. The world had a lesser reliance on <u>fossil fuels.</u>

We can learn to do that again too!

Small, simple changes can add up to big results, and making a difference starts with us!

We can all become <u>climate-neutral</u>!

Are you ready to begin your journey as a <u>climate activist</u>?

So what can you do?

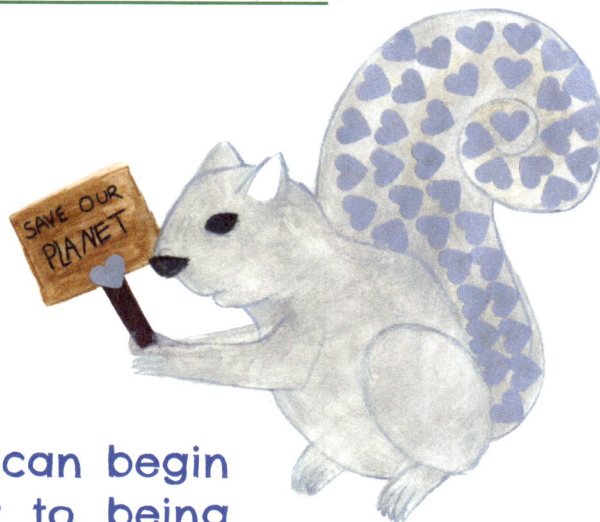

There are three areas that you can begin with that will bring you closer to being <u>climate-neutral</u>.

1. Eliminate your <u>food waste</u> from the garbage by <u>composting</u> it.

2. Stop using <u>single-use</u> items.

3. Start eating more plant-based foods (fruits, vegetables, nuts, beans) because they have the lowest <u>food carbon footprint</u>.

Here are some more ideas:

- Share this book with your friends, family, and classmates.

- Organize a toy swap, book swap, and/or clothing swap event at your school or in your community.

- Take only a serving of food that you know you can eat. You can always go back for seconds if you are still hungry.

- Take a sewing class or ask an adult to help you learn. You can then repair small holes in your clothes or add fun patches so they can still be worn.

- Donate toys and clothes that you no longer need prior to buying new ones.

- Help by turning off the lights and water when not in use to save energy.

- Ask an adult to help you build a small garden. You can grow your own herbs indoors all year or have your favorite fruits and vegetables grown in your own yard.

- Time yourself when you take a shower and try to get it to five minutes or less. That will help to save water.

And even more ideas!

- In the winter, put on a warm layer, including cozy socks, and ask your family to turn down the heat.

- Always bring your own refillable water bottle from home.

- Eat more vegetables. Vegetables are not only good for you, they are also the <u>lowest</u> <u>carbon emission food</u>.

- When you make a wish list, try asking for experiences instead of presents. Maybe you can go for a hike and picnic with a friend instead of getting another toy.

- Try and stay away from wrapping paper and bows unless you can make them from recycled or reused materials. Wrapping paper and bows are <u>single-use</u> items.

Can you think of even more ways that you can become a climate activist and reduce your carbon emissions?

For more ideas check out:
www.TheCarbonEmissionsProject.com

Here are a few <u>climate</u> actions that are happening worldwide...

Paris Agreement:

Also called the Paris Accord

Signed in 2015, it is the world's first treaty on <u>climate</u> change.

The Paris Agreement set the goal to limit the global temperature increase to 1.5 degrees <u>Celsius</u>. Countries around the world agreed to help each other by sharing information and support.

The 17 Sustainable Development Goals:

(SDGs)

Adopted by the <u>United Nations</u> in 2015, with the goal of achieving them all by 2030, so that no one gets left behind.

1. No <u>poverty</u>

2. Zero hunger

3. Good health and well-being

4. Quality education

5. Gender equality

6. Clean water and sanitation

7. Affordable and clean energy

8. Decent work and economic growth

9. Industry, innovation, and <u>infrastructure</u>

10. Reduced inequalities

11. <u>Sustainable</u> cities and communities

12. Responsible consumption and production

13. <u>Climate</u> action

14. Life below water

15. Life on land

16. Peace, justice, and strong institutions

17. Partnerships for the goals

THE NINE PLANETARY BOUNDARIES:

Scientists created these limits (boundaries) to keep the Earth safe. The boundaries define a set of rules that we can follow to keep the planet healthy.

Just like the rules that you have to follow at home and school, scientists hope that we can follow their nine important rules to keep our planet healthy and happy.

1. <u>Climate</u> change - don't use too many fossil-fuels because the Earth will get too hot.

2. <u>Biodiversity</u> loss - protect all the different kinds of plants and animals.

3. Freshwater use - protect the water in our rivers and lakes.

4. Ocean <u>acidification</u> - (as-id-if-fuh-kay-shun) protect our oceans and sea creatures.

5. Ozone layer <u>depletion</u> - protect the layer that shields us from the sun's rays.

6. Land system use - protect the forest and land, and keep it in its natural state.

7. Nutrient pollution - be careful not to use too many chemicals on our land and in our water.

8. <u>Atmospheric aerosol</u> loading - be careful about what we put into our air.

9. <u>Novel entities</u> - be careful not to introduce new materials that are harmful to the environment.

CERTIFICATE OF AWARD

This certificate celebrates the day that I became a

CLIMATE ACTIVIST

Name _____   Date _____

Congratulations
Kimberly

Luna

Certificate of <u>Climate Activism</u>. Represents my commitment to: lowering <u>carbon emissions</u>, eliminating <u>single-use</u> items, and diverting <u>food waste</u>.

ASK ME HOW YOU CAN HELP MAKE OUR PLANET MORE HEALTHY!

Just for fun!

NO-BAKE LOW-CARBON HEALTHY TREAT RECIPE

Almond Joy Protein Bites

Mix all ingredients in a bowl. It may be best to use your hands. Make sure you wash your hands well with soap and water before mixing the batter. Once the ingredients are mixed, use a small scoop to form the batter into small balls. You can use your hands to help shape the balls. The batter will make approximately 24 protein bites.

Ingredients:

1/2 cup gluten free, organic old fashioned oats

1/3 cup almond butter

1/4 cup maple syrup

1/4 cup ground flax seeds

2 tablespoons plant-based chocolate protein powder

1/3 cup unsweetened shredded coconut

1/2 cup vegan chocolate chips

Enjoy your tasty treats with a friend!

GLOSSARY

- Absorption: (absorb) when one substance takes up another, like a sponge cleaning up a puddle of water.

- Acidification: when something like water and soil has too much acid in it.

- Adapt: adjust or modify.

- Aerosol: particles in the atmosphere that are so small they are not visible. When we talk about the environment, they can have a significant impact on the health of the planet.

- Agricultural: another word for farming activities.

- Anthropogenic: caused by human beings.

- Atmospheric: relating to or occurring in the layer of gas (atmosphere) surrounding Earth.

- Beneficial: helpful, useful, good.

- Biodiversity: containing a variety of life.

- Bog: a swampy ground made up of decomposing plants and moss.

- Brainstorm: a fun and creative way to come up with solutions, usually a group activity that encourages participants to share ideas as soon as they think of them.

- Carbon Dioxide (CO2): one of the primary greenhouse gases (GHG). You may have heard of CO2 before if you studied photosynthesis. We exhale CO2, and plants "breathe" it in (through their leaves) and turn it into oxygen for us.

- Carbon emissions: the releasing of CO_2 into the atmosphere.

- Carbon footprint: the amount of emissions that we generate through our daily activities in life.

- Celsius: just like Fahrenheit, measures temperature, it just uses different units of measurement. Celsius is used in almost every country in the world, as well as by most scientists. The United States, however, uses Fahrenheit. As an example, you may know that water freezes at 32 degrees Fahrenheit. In Celsius, water freezes at 0 degrees.

- Climate: long-term atmospheric trends of a large region.

- Climate activism team: a group of people working together to take action on behalf of the planet in an effort to defeat climate change.

- Climate activist: anyone working to protect the planet's health.

- Climate neutral: an equal balance of emissions and absorption.

- Climate overshoot: the period of time after the Earth's temperatures increase past the 1.5 degree Celsius mark and are stabilized, but the effects of climate change are still felt. This is different from Earth Overshoot Day - which marks the calendar day each year when Earth or a particular country has used up the natural resources for that year. In 2025, the United States used up its annual natural resources up by March 13.

- Carbon footprint of food: see; Food carbon emission.

- Carbon sinks: absorb carbon; examples are our forests, oceans, and soil.

- Collective: a group of people acting together.

- Composting: a process that turns food waste into a material that can be used to enrich the health of the soil.

- Conclusion: a result based on observation.

- Consumerism: the idea that buying more things is important for a good life.

- Containing: to hold something within.

- Convenience: the thing that is quickest and easiest.

- Deforestation: the destruction of an area of trees.

- Depletion: decreasing the amount of something - like a natural resource.

- Earth Day: happens every year on April 22nd. It is a time to raise awareness about protecting the planet. Check with your town to see if there are events that you can participate in, or maybe YOU can organize an Earth Day event!

- Effect: to cause something to happen.

- Emissions: released into the environment, often referring to a type of gas.

- Enchanted: to be filled with delight.

- Environmental: focused on protecting nature.

- Finite: having a limit.

- Food carbon emission: the amount of greenhouse gas (GHG) that is created by the production, transportation, handling, distribution, preparation, and disposal of food. Also called the carbon footprint of food.

- Food waste: food that is not eaten and gets thrown in the garbage.

- Fossil fuels: coal, petroleum (oil), and natural gas.

- Gigaton: a measurement of one billion metric tons. It's a lot!

- Global warming: is the increase Earth's temperature.

- Grains: seeds of grass that are grown for food, like wheat, rice, and oats.

- Greenhouse: a building with a glass roof and glass sides so that it lets in the sun. It can be used to grow plants in a protected and warm place.

- Greenhouse effect: the gases in the atmosphere that hold the sun's heat close to Earth.

- Greenhouse gases (GHG): any gas in the atmosphere that traps heat, primarily carbon dioxide, methane, and nitrous oxide.

- Illustrator: an artist who creates pictures for a book.

- Infrastructure: the systems that serve a community, for example - schools, roads, and bridges.

- Insights: a deep understanding.

- Inspiration: when you have a great creative idea that pops into your head.

- Inspires: to excite and interest so that you want to take action.

- Landfill: the place where our garbage goes. Sometimes it is buried underground, and other times it is left as a big pile. Yuck!

- Legumes: any plant that bears its fruit inside a pod, like lentils, peas, broad beans, chickpeas, soybeans, and peanuts.

- Marveled: amazed, surprised in the best way.

- Methane: one of the greenhouse gases (GHG). It comes from landfills and farming, as well as other sources.

- Microplastics: tiny pieces of plastic that break apart and remain in the environment where they can cause damage by polluting soil and water.

- Nitrous Oxide: a greenhouse gas (GHG) produced by agricultural activities.

- Novel entities: things created by human beings that negatively affect the environment. Microplastics are an example.

- Plankton: microscopic plants or animals that drift in water (or the guy from Sponge Bob).

- Poverty: not having enough money to buy things like food and housing.

- Reasonable: fair, with good judgment.

- Sacrifice: destroy, or the act of giving something up.

- Single-use: something that you use only once and then throw away.

- Specific: exact and detailed.

- Stabilized: to be fixed, steady, firm.

- Sustainable: maintained, continued. In regard to climate change, sustainable means protecting resources for future generations.

- United Nations: An organization of countries from all over the world that work together to solve problems like climate change.

- Values: the beliefs that determine what is most important.

- Weather: the small temporary changes, like rain or snow, happening in a small area, like your town or state.

-